Some Say Tomato

poems

Copyright August 1993
Mariflo Stephens

ISBN 0-9638892-0-6

All rights reserved. No significant portion of this book may be reproduced without permission of the publisher, writers, and copyright-holders. Short segments for review, reference or discussion purposes is granted hereby, with appropriate credit line.

Several poems have been previously published and are reprinted with permission as identified in the Acknowledgments.

Correspondence may be directed to
Mariflo Stephens
108 2nd Street, SW, No. 5
Charlottesville, Virginia 22902

Cover photo is courtesy of
Tom Cogill
Charlottesville, Virginia

Printing, type, and graphics by
Papercraft Printing and Design Company
114 Old Preston Avenue, Charlottesville, Virginia U.S.A.

Acknowledgements

My thanks to the editors and publishers of the following publications, in which these poems first appeared.

Yellow House on the Corner (Carnegie-Melon University Press): The Secret Garden, Adolescence—III (reprint by permission of the author only)

House, Bridge, Fountain, Gate (Viking): The Mummies (reprint by permission of the author only)

Ms.: Preparing for Winter

Stalking the Florida Panther (The Word Works): The Tomato Packing Plant Line

Life Scribes: Caledonia Winter

Plain Air (University Presses of Florida): The Okra Flower

Ploughshares: Cancion de Segadores

Cellar (Alderman Press): Farm Poems

A Quarter Turn (The Sheep Meadow Press): To Keith

Walking Distance (University of Pittsburgh Press): Produce

Selected Odes of Pablo Neruda (University of California Press); Ode to Tomatoes

The Nature of Yearning (Gibbs Smith, Peregrine Smith): The Nature of Yearning

Rowing Past Eden (Nightshade Press): Cash Crop

My gratitude to the poets who read this collection, especially to Marilyn Kallet, whose fine hand at editing and ordering straightened out the seams, and thanks to M.A.D. for enthusiasm and support. Thanks also to Fred Heblich whose typesetting skills and knowledge of printing came in handy.

Table of Contents

Acknowledgements

Contents

Introduction

1 ... Mariflo Stephens ... *The Tomatoes Arrive in 1993, During the Storm of the Century*
2 ... Rita Dove ... *The Secret Garden*
3 ... Maxine Kumin ... *The Mummies*
4 ... William Stafford ... *Voices in the Garden*
5 ... Marilyn Kallet ... *Alone and Eating, Valentine's Day*
6 ... Connie J. Green ... *Maybe, Tomatoes*
7 ... Al Young ... *Seeing Red*
8 ... Helen Fahrbach ... *Harvest*
9 ... Bobby Caudle Rogers ... *My Father's Kitchen*
10 ... Helen Fahrbach ... *Legacy*
11 ... Rusty McKenzie ... *Sleep*
12 ... Nila Jean Robinson ... *Indian Summer*
14 ... Debra Allbery ... *Produce*
15 ... Laurel Mills ... *Preparing for Winter*
16 ... Mariflo Stephens ... *December Tomato*
17 ... Enid Shomer ... *The Tomato Packing Plant Line*
18 ... Joan Z. Rough ... *Caledonia Winter*
20 ... Amy Spence ... *Advice to Gardeners*
21 ... Ellen Kort ... *Tomato Fight*
22 ... Marilyn Kallet ... *Persephone's Prize Tomato*
23 ... Ellen Kort ... *Something Red*
24 ... Nancy Kober ... *Alf Landon Makes A Whistle Stop in Penn-Shaft*

Bad Rap . . . Jane and Isabel Heblich . . . 28
Boing! Tomato . . . Heather Miriam Gross . . . 29
A Tomato I Once Knew . . . Emily Morris . . . 30
A Tomato from Grandma . . . Blaise Zerega . . . 31
Identity Crisis . . . Al Stephens . . . 32
Family Garden . . . John McCutcheon . . . 33
Caution: Pommes D'Amour . . . Stephen Margulies . . . 34
Getting from the Garden . . . Stephen Margulies . . . 35
Juice Atonement . . . Tonya Miller . . . 36
Lust . . . Kristen Staby Rembold . . . 37
Ode to Tomatoes . . . Pablo Neruda . . . 38
The Okra Flower . . . Michael McFee . . . 41
Adolescence—III . . . Rita Dove . . . 42
Farm Poems . . . Lisa Russ Spaar . . . 43
All Tomatoes . . . Connie J. Green . . . 44
After Too Much Sanity . . . Susan Ludvigson . . . 45
Harvest . . . Sydney Blair . . . 46
Putting Up Tomatoes . . . Amy Spence . . . 47
Okra . . . Shirley Anders . . . 48
Poor Song . . . Stephen Margulies . . . 49
The Manikin Leaves . . . Jane Caris . . . 50
For My Daughter . . . Judy Longley . . . 52
To Keith . . . Debra Nystrom . . . 53
Cancion de Segadores . . . Antonia Quintana Pigno . . . 54
Ephemeral Landscapes . . . E. A. Gehman . . . 58
A Letter to Michael McFee . . . Shirley Anders . . . 60
The Nature of Yearning . . . David Huddle . . . 62
Cash Crop . . . Judy Longley . . . 65

Introduction

The enormous pleasure I've felt collecting a book of poetry like this was diminished one day by a friend's reaction after I told her one of the many anecdotes I've gathered about tomatoes and poetry since I decided to marry the two between the pages of a book.

"Oh," she said, "And are you going to write about that in the introduction?"

Write? Introduction? Suddenly all this pleasure took on the feel of work. What *is it* about tomatoes? Someone is bound to ask me. What will I say in the introduction?

I could say that poetry in American has been so neglected, so misunderstood that to work in that area today practically represents an act of courage. I broadcast poetry over the student radio station at the University of Virginia, and I've often quoted this statistic: ". . . as many as 14,000 people have crowded into a Moscow sports stadium to hear (Voznesensky) recite his works." Then I talk about how hard it is sometimes to get an audience for poetry in America.

I could write about how our current United States president, William Jefferson Clinton, elevated the cause of poetry by commissioning a poem for his inauguration earlier this year.

Or I could simply describe how the project began. One summer evening when I was pregnant and hot, I made myself a promise, not really believing I'd keep it. It was after a small dinner party at my friend Judy Longley's stone house in the country and I sat in her living room wishing for the glass of wine my pregnancy denied me. Judy asked if she could share a recent poem and began reading "Cash Crop." Listening to it, I was lulled into a state of transformation more complete than any glass of wine can manage. I was a child again, rummaging among tomato vines for some magic.

Then another poet, Shirley Anders, visiting from out of state, announced that she too had a poem whose central image was that of a tomato. Me too, I thought, but did not admit aloud. Though I am a story writer, I've composed a dozen or so poems in my adult years, and two of them (now three) contain tomatoes. So I promised myself that one day I would take on the job of putting out a collection of poems of, about, or around the tomato.

Several of the poems in this collection link tomatoes to healing and more than one connects this fruit with desire. Historically, the tomato was, after all, called a love-apple and considered poison.

The birth and death metaphors alive in any garden need no explanation from me. David Huddle's fine poem, "The Nature of Yearning" pretty well sums it up, with "as one perfect trumpet note still / this clarity this concerto suggests / the coming death of tomato vines..."

Love and homegrown tomatoes—the only things that money can't buy. Anything further I could say about tomatoes has been expressed in this volume and elsewhere in the culture. Like in Guy Clark's song, "Homegrown Tomatoes"—"Put 'em on the side, put 'em in the middle. Put homegrown tomatoes on a hotcake griddle." Or in Peter Taylor's short story " The Oracle at Stoneleigh Court," also published in a collection of the same name this current year of 1993. *When we were talking a while that day among the upside-down tomato plants, suddenly I found myself stepping forward and then leaning among those upside—down plants to embrace for the first time this girl I knew I had fallen in love with.*

Still I worried over what to say about tomatoes in this introduction, worried so hard I delayed turning in the disc to the printer. But I'm glad. Marilyn Kallet sent my letter about tomato poems to Al Young and William Stafford, who composed "Voices in the Garden," a poem for this collection. His generosity inspired me to write "The Tomatoes Arrive in 1993, During the Storm of the Century." Marilyn convinced me that this poem was a dedicatory poem and should be at the beginning and even though it embarrasses me to put it first, I'm going to abide by her advice.

Meanwhile, Al Young ruminated on the tomato and the very day I set out to finish this book and put it out of my hands and mind, he faxed me "Seeing Red." It took Al Young to see through the eyes of a tomato, I thought, and then realized that, months before, on an opposite coast, my nephew Al Stephens had also considered life on the vine.

— Mariflo Stephens
Summer 1993, Charlottesville, Virginia

In Memoriam

William Edgar Stafford

January 17, 1914 – August 28, 1993

Some Say Tomato

Christmas 1993

Best wishes —

Mariflo Stephens

for my mother, Isabell Dean Johnstone Stephens
October 15, 1916 — December 9, 1974

and for my aunt, Flora Virginia Stephens
March 6, 1918 — December 9, 1991

The Tomatoes Arrive in 1993, During the Storm of the Century

Maybe this will heal me—
a mid-winter thought,
summer avalanche.
Spilled onto snow
red, round faces shoot
towards Virginia from
San Francisco, Lake Oswego, Oregon
arrive by the brown UPS truck
the denim-suited mailman
whose sad squint reminds me
of mourning you.

I tried to be the perfect daughter,
compensate for the son lost to polio
six weeks before I arrived.
The doctor warned you,
"The polio will claim this baby too, if not
that, then the German measles you had, early on."
You carried me sick with grief,
dreading another death.
Oh Mother, remember, the doctor held me
in his palm like a late tomato
regarding this puzzle, me,
mysteriously whole.

— Mariflo Stephens

The Secret Garden

I was ill, lying on my bed of old papers,
when you came with white rabbits in your arms;
and the doves scattered upwards, flying to mothers,
and the snails sighed under their baggage of stone . . .

Now your tongue grows like celery between us:
Because of our love-cries, cabbage darkens its nest;
cauliflower thinks of her pale, plump children
and turns greenish white in a light like the ocean's.

I was sick, fainting in the smell of teabags,
when you came with tomatoes, a good poetry.
I am being wooed. I am being conquered
by a cliff of limestone that leaves chalk on my breasts.

— Rita Dove

The Mummies

Two nights running I was out there
in orange moonlight with old bedsheets
and a stack of summered-over Sunday papers
tucking up the tomatoes while the peppers
whimpered and went under and the radishes
dug in with their dewclaws and all over
the field the goldenrod blackened
and fell down like Napoleon's army.

This morning they're still at it, my tomatoes
making marbles, making more of those little
green volunteers that you can rattle
all winter in a coat pocket, like fingers.
But today on the lip of the solstice
I will pull them, one hundred
big blind greenies. I will stand them
in white rows in the root cellar
wrapped one by one
in the terrible headlines.

— *Maxine Kumin*

Voices in the Garden

"Pretty soon," spring flowers whisper,
"summer will come,
pretty soon now."

"After while," trills the Queen Anne's lace,
"It will be autumn,
after while."

"Right now," headlong tomatoes boom,
"summer needs us all,
right now."

— *William Stafford*

Alone and Eating, Valentine's Day
(for William Stafford)

Don't try to distract me
with tomato juice
from the bitter, fermenting reds of love!

Cold news. It's February,
the garden's just a plot,
a plot.

Give me something to hold
against summer's promissary—
something smooth

as that ersatz fruit,
firm
tight

    like
        the traveling sun's slow heat
wrapped in satin,

fleshy
and predictable,
Bill.

When love's not,
food
is.

— Marilyn Kallet

Maybe, Tomatoes

if the vines mature
if the caterpillars don't get them

if we water, sucker, feed
if we pick and preserve

maybe, tomatoes
    thin sliced on sandwiches
    chunked into salads
        peeled and whole
        juiced and sauced
            stewed
                pickled
                    stuffed

                              – *Connie J. Green*

Seeing Red

You always seem to get it all wrong about me.
Just like in the days back when
you thought you'd up and die if you chomped me down,
so you ate my leaves instead and wound up dead.
Now you think it's OK to keep me from dying,
so you actually poison me through irradiation.
Where's your imagination? Where's the spirit
of the Aztecs, who grew me to death, named me tamatl,
and loved me for the very fruity berry that I am?
From Plato to NATO, the vegetable consciousness
of Western Civilization mineralizes its own
pockets; oil-cloth pockets so you can steal soup.
That growers in this nation would stoop
to chemotherapy to give me greater so-called shelf-
life may hold the answer to cancer, but it doesn't
do a thing for me. I like to salt and spice
your mouth up, then seed it all red with zesty juice
and yellow-green after-thoughts like the bright
ting-a-ling of love. You hear what I'm saying?
You hear what I'm telling you? Rather than right
those ancient wrong notions, you've motioned them on.
Like edible street gangsters now, rain or shine,
we don't die, we multiply. Tell Henry Heinz
we tumorless tomatoes constantly see the best minds
of our generations goosed, juiced, and pissed.

— Al Young

Harvest

Drone of bees, whir of grasshoppers,
pungent odor from tomato vines
and our hands stained strangely yellow
as the crimson harvest
fills brown grocery bags.
When afternoon light
sifts through oak trees
we climb the slope to the cabin.

Later in the kitchen
we peel those wet red hearts,
slip them into a kettle
to make spaghetti sauce
and sharp tomato fragrance
takes me back to childhood kitchens
where pickles, jams and chili
scented the house like love.

— *Helen Fahrbach*

My Father's Kitchen

Before the twelve-hour drive I sit looking
at the pine walls of my father's kitchen
stained to a red the color of the sunrise
just beginning down the street.
Elliptical drops of tap water
fall one at a time from the chromed faucet
and burst against the steel of the sink.
Smoke twists from the coffee that is too hot
to taste. A row of still-green tomatoes,
pulled from the vine too soon, has been left
on the window sill to ripen. My father
stood here yesterday, arms crossed, looking
to where his plot of tomato plants bears
heavying fruit in the far corner of the property,
staked and tied fast with strips of fabric.
He told me that the soil of this hill is no good
for much else but will keep a tomato vine full
of yellow blooms till the summer abates.
I am close to believing he had more to say,
but if he did, it is his business; mine lies
at the eastward end of a long day's drive.
He will come back in a few minutes, taking
off from work to watch me leave again,
and ask if I can stay longer next time,
ask when it will be. The kitchen window
is full of fingerprints. In the spring, I will say.
The pine cones in the back yard lie scattered,
already wanting the rains and fits of snowfall
the intervening months must bring.
I remember when lightning split
the pine tree across from our old house—
how the splintered wood was white for weeks.

— Bobby Caudle Rogers

Legacy

Opening a new garden book,
I think of my father
planning his garden in February.
He sat at the kitchen table, surrounded
by seed catalogs, his chapped hands
pausing at runner beans, lemon lilies.

I remember finding his notebooks,
lined paper carrying his records:
temperatures, rainfall, dates when
scilla or crocus first appeared,
yield of tomato vines.

Dreaming ahead to summer
I think of those times before dusk
when he finished the last weeding,
brought his gifts: sweet blue grapes,
one red rose, his hands dark with earth,
the fragrance of sun around his head.

— *Helen Fahrbach*

Sleep

Sleep wrestles in fuzzy nightpockets,
slips under covers, mute as moonpuddles
on speckled linoleum. Sleep comes in waves
some nights, like wind in the pillow, as if
hungry sharks were hunting the corners of dreams,
too lazy to slip the latch, free the horses.

I sleep beneath seven purple globes
under a blanket of lemon-summer fog,
undisturbed until Monday when your mother calls,
and then I wake scratching mosquito bites.
As if it mattered.

If all sleepers got together and voted
for a Democrat, who would win?
Sleep comes under the heading of useless activity
until dreams—land of toadstool
and sandburr, poems tattooed on a white
dolphin, flags over the chicken coop.

Where lovers are masked, ashes clog
the freeway and coconuts run naked
through the turtle farm. If sour apples
fly into dreams, it probably means your mother
nursed you in church during Lent.

The best sleep rides the midnight train
to Hope Town, wears one chocolate-covered earring
and a fur coat with tomato-seed buttons,
softly whistles The Muskrat Ramble.

— Rusty McKenzie

Indian Summer

Warm afternoons in Indian summer
my mother escorts tomatoes
from the garden to sunny ledges,
from ledge to canning kitchen.

Together we pick each with a twist,
gently and without insult
rub off the platformed steam which crowns
each fruit like a green star of anise.

We queue them up on eighteen inch plank,
lay them out over upturned barrels of Wolf's Head Oil
before a west window in Father's garage;
they smell of thyme and vinegar.

Harvest sun distills the green spots,
sheltered by sunken earth and heavy, spiked leaves.
Light saturates the dense produce,
swells it almost to bursting.

The dimpled acid fruit heats even redder
in a familiar heavy kettle that occupies half the stove.
Mother plunges them scalding into the stainless steel sink,
buries her wrists in the cold water.

Cradling the pulpy balls in heavy lined palms,
she rocks them, softly blanched, between springy fingers.
Her working hand and the wedding-ringed hand
shuffle off skin in clinging shreds.

Flayed red bodies settle denuded.
Tattered hides float, oscillate as kelp in current.
Bloody trophies cloak the three moles
marking a straight line between thumb and forefinger.

One tatter wraps itself around her wedding ring,
hides the purple blue line etched by time and weight,
stained by fifty Midwest summers
into the swollen rebellious flesh.

This is an old memory; twenty more summers have gone.
Red orange jars from her last harvest still crowd
pantry shelves with her pickles and peaches.
When they have gone, her hands have died.

On a warm afternoon in Indian summer,
I will buy a bushel of tomatoes, curse whose slackness
left one stemmed to bruise its fellows,
carry each to the sunniest window I can claim.

On a day good for work, I will lean
my forearms on a cold bright sink,
and look down at my hands,
fumbling with the ripe red bodies.

— *Nila Jean Robinson*

Produce

No mountains or ocean, but we had orchards
in northwestern Ohio, roadside stands
telling what time of summer: strawberries,
corn, apples—and festivals to parade
the crops, a Cherry Queen, a Sauerkraut Dance.
Somebody would block off a street in town,
put up beer tents and a tilt-a-whirl.

Our first jobs were picking berries.
We'd ride out early in the back of a pickup—
kids my age, and migrants, and old men
we called bums in sour flannel shirts
smash-stained with blueberries, blackberries,
raspberries. Every fall we'd see them
stumbling along the tracks, leaving town.

Vacationland, the signs said, from here to Lake Erie.
When relatives drove up we took them to see
The Blue Hole, a fenced-in bottomless pit
of water we paid to toss pennies into—
or Prehistoric Forest, where, issued machine guns,
we rode a toy train among life-sized replicas
of brontosaurus and triceratops.

In winter the beanfield behind our house
would freeze over, and I would skate across it
alone late evenings, sometimes tripping
over stubble frozen above the ice.
In spring the fields turned up arrowheads, bones.
Those slow-plowing glaciers left it clean and flat here,
scraping away or pushing underground what was before them.

— *Debra Allbery*

Preparing for Winter

We gather apples from under the tree, tossing away
ones squirrels have nibbled, swatting at bees
as we fill the wheelbarrow with red fruit
smelling of cider. Yellow leaves from the basswood
spill over grass. Soon we'll suck them into bags
with our lawn tractor, add them to the compost heap.

How I love transformations: sunsets the cooling air
smears across sky, crimson as sugar maple. Slowly,
freeze begins at edges, shriveling tomatoes, browning
the mums. What will be left to say when first snow
finally falls—that we've known all along
it was there, haloed by a thin layer of sky,
just waiting to sieve its way through?

Some truths we hoard without tongue to articulate.
There isn't a word complete enough to describe
the crisp vapor we name frost—the way it covers
everything, how even wind cannot move it, how it seals
the mouth of the last peace rose. Then, as morning
wears on, sun meddles, the only light that knows
how to weave between cold granules, thin them out
so even dead basswood leaves seem to breathe again.

It has taken all my life not to dread pendulous snow
that will bury the few apples remaining, leaves left
unraked, the last cherry tomato on the vine.
It has taken all my life, in this world of strangers,
to find words to say: there is a woman I love.

— *Laurel Mills*

December Tomato, 1987

    for my husband

Before Campbell Soup closed the Camden plant,
boys in blue jeans picked for a day's pay,
fed the truck farms that line the New Jersey highway.
I imagine you, your arms leaden,
sucking sunshine, growing dreams
before tomato soup
is canned into obscurity, Warholed into a giant joke,
what you must have planned.

Stashed in brown paper in the dank pantry
the September the tomatoes were gone,
at Christmas the green had ripened into red
I savored it in secret
remembering your hand deep
in the warm earth,
what we've made.

— Mariflo Stephens

The Tomato Packing Plant Line

Bumped and rolling jovially
down the conveyor the tomatoes
dance in   a press of faces
that shine on their skins like smiles
the stem ends chipper as cowlicks.

Young women remove the mistakes—
harelips   two-headed one   gashed ones
with papery crosshatched scars.
Tiny ones too are removed
to be juiced with the freaks.

At the far end hemmed in by boxes
the old women sort the tomatoes
the largest and perfect ones first.
Their hands   like their eyes
know the swell before ripeness.
It is something they flaunted
on Fridays   a gust that inflated
box-pleated skirts into bells
as they stepped into dusk
hands washed clean of tomatoes
which did not survive
their ripeness.

— Enid Shomer

Caledonia Winter

Paperweight swirls build
knee-deep drifts by the door.
At dawn I carry steaming buckets
to sheep breaking a path
through frozen light;
drop sweet bales from the haymow
to hungry beasts thankful
for shelter against the cold;
collect warm eggs from hens
huddled in the corner of the coop;
carry splits of oak and ash to the stove
slowly warming the kitchen
where my children squabble over
Matchbox trucks
and whose turn it is to do what.
Toast rinds and spilled oatmeal
grace the table.
Later I'll banish them
to make angels in their own image;
build forts for frozen knights
armed with wilted carrot noses.
I'll watch from the window
kneading in fevered winter frustration,
while evening grosbeaks mob the feeder
stoking their furnaces to warm them
through the coming night.

I'll sit in gray afternoon light,
Seed catalogues before me,
ordering lush tomatoes,
pole beans and pink melons;
yearn for spring asparagus,
strawberries for pie;
dreams that warm me
as I trudge sloshing water to sheep
jostling to be first in line
for the evening feed.

— Joan Z. Rough

Advice To Gardeners

Wait for the garden.
Nothing weathered is barren.
Fallow emanates a solicitous light.

Intent on color
we planted
row upon orderly row
of scentless roses.

See, the geraniums turn red faces,
the sunflowers droop away,
and the renegade vines of tomatoes climb
circuitous, snaking, over the gate

where the insects gather and chirp
This is the beginning
This is your fate

— Amy Spence

Tomato Fight

I don't know what started it
Maybe it was the heat of late summer
or the fact that we were standing

in the garden   sucking the warm blood
of tomatoes   pretending to be
tribal warriors eating the hearts

of our enemies   taking in their strength
and power   I think at first we were
dancing   the urge of roots   a need

for brightness   red-winged dance
of skin and flesh   ripening seeds
bursting inside the belly of tomatoes

inside the dark and secret heart
You threw one then   a crimson slash
against my white blouse   and I commended

myself for my quick far-flung response
We couldn't stop   It was delicious
all that sweetness   the scarlet nectar

the musky tissued flesh   unhinged from
its skin   how the red body so easily splits
in two   opens its strong young thighs

                         — *Ellen Kort*

Persephone's Prize Tomato

Persephone's patch
harbored
plumpness—

on a sanguine
tip
she plucked,

bit,
& savoring it
summerlong

would not
go down.

No season off
for Persephone?
Stuck

with her mama—
like pasta
in cold sauce.

— Marilyn Kallet

Someting Red

Who knows
what tomatoes dream
bedded
in their underground den
stretching naked
a slow hunger
moving upward
sweet and singing
The Sucking sound
A leap of green
raises itself
toward sun
into quivering
swing of
blossoms
into berries
that turn
their small
hard bodies
like a sudden
clench of
fisted hands
Finally the air
turns crimson
birth-damp
morsels of red
expand
like balloons
Shine
bright
lanterns
in the day's
half-light

— *Ellen Kort*

Alf Landon Makes
A Whistle Stop in Penn-Shaft

Everyone was blessed (as Mum saw it) with bumper yields all summer
and well into fall. Plump balls of sun brightened our cellars,
crowded our pantries, perked up our ham loaf, our *golumki,*
perfumed the black cookpots of Italian miners' wives. Red armies

spilled from huckster wagons, avalanching down brick alleys,
sacrificing their slurpy guts to the tires of the coal trucks. They
were omnipresent, these vegetables, like FDR's voice on the radio.
Jobless uncles juggled them between innings of fast-pitch softball.

Hoboes chomped them like peaches on our back porch, courtesy of our
generous but guarded mum. Remember her ministry of liniment and
horehound drops? How she wiped her face with her red-speckled apron?
How she doled out movie money from a red dimestore teapot shaped like—

well, I said they were everywhere. Insinuating fuzzy stalks
through the neighbor's fence. Escaping their flimsy prisons
by night. Infiltrating the railroads (where Dad sold tickets at
Penn-Shaft station), lending a delicate blush to a dining car rarebit,

a burnished glow to an aspic. Their shapes suggested by the girdled
plenty of a shopgirl's bottom, their name given slangy new meanings
by George Raft lookalikes. Smell them simmering in the vast kettles
of the Heinz plant, an essential ingredient for fourteen of the

"57" varieties—which, incidentally, could just as easily have been "59"
or "62"—old H.J. Heinz (whom Father admired greatly) picked that
number for its psychological resonance, not its accuracy: an auspicious
conjunction of bankerly "five" with luck-be-a-lady "seven." To some

clever secretary fell the challenge of revising the product list
so it totaled exactly "57." A task made trickier by successful
diversifications. Which she accomplished, it seems, by double counting
Boston-style and regular-style baked beans and lumping all the strained

baby foods into a single variety. Did I mention that Father was
a Republican, an Alf Landon man? That his shirts, though pummeled
to gauze by Mum's vigorous scrubbings, were impeccably creased and
starched? That he wrote in a forceful yet beautiful hand, crossing

his sevens in the Old World manner? That he could handle most
emergencies with a penknife, some twine, and a paperclip? Five years'
formal schooling, correspondence courses in bookkeeping, the Transcen-
dentalists, Problems of Democracy. But let's not forget those poor

hoboes relegated to the back porch, rumpled and sore from sleeping in
cold coke ovens—decent men mostly, just down on their luck; though
probably not as earnest as Gary Cooper's John Doe, certainly not as glib
as William Powell's Godfrey. But hard-working men nonetheless. Except

some of them were reds. Take those fellas who played Rook outside
the Feed and Flour. Subversive words were overheard. By summer's end
we'd eaten enough to last a lifetime. That acidy taste overpowering
our soups, souring our sandwiches, polluting every burp. What we

wouldn't give for some mushrooms, a dozen ears of sweet corn.
With boatloads of big ones still ripening on the vines, green bombs
still lurking behind the leaves—sometimes you just wanted to rip 'em out
by the roots. On Columbus Day a blast swept down from the lakes

and froze them. According to Father, the Landon retinue was not
scheduled to stop at Penn-Shaft. But still the people turned out.
Pit bosses, ladies in rain bonnets, mill hunks, farmers from their
fields, the woman with the cantaloupe-sized goiter—all jamming

the platform, hoping for a glimpse of the man who might be President.
(Mathematically it was still possible, what with the electoral college.)
Father slipped from his cage, and imagine his joy when the air brakes
squealed, the engine slowed with a bullsnort, and look, there's Alf,

waving from the caboose balcony! Gracious, whispers Mum, he looks
gaunt. And so he does, a ghost in a black homburg. Speech, speech!
shouts Father. The Governor obliges. A very commonsensical statement,
we later agreed—government as umpire in the economic ballgame,

not overpaid star player. Nobody else sees Rudy Zunka's uncle (an
anarchist, a believer in the doctrine of propaganda through deed)
stomp out his Lucky Strike, slide his hand in his coat pocket, coddle
a suspicious bulge, assume a sharpshooter's squint, aim, and fire off

the most gigantic, swollen, juicy—barely held together by a thin membrane—
TOMATO, the choicest blooddrop from a dying stalk. Splat! Right on
the noggin! Calmly, Alf draws a hanky from his breast pocket,
mops the trickles from his cheek, and with the peaceful eyes of Jesus

smiles wanly at the cackling mob. Someone's pretty little daughter
shoots a pea. You won't find this in anyone's memoirs, or the morgues
of the local papers. Maybe the crowd wasn't so huge after all, maybe the
reporters had gone chasing after Roosevelt to Cleveland. That such

hooliganism could happen in our town, to a fine American—a *progressive*
for goodness sakes!—well, Father didn't know what else to say. Dogs lost
to rabies, a child to scarlet fever, a brother to a mine collapse, crops
to blight, jobs to economic cycles—these things were by no means foreign

to Father. He mourned them, of course, then accepted them as the will
of God. Study, pray, work harder—that's what men of character did.
Meanwhile a rainbow of mason jars bided time in the cellar: anemic
globules floated behind glass, red sauces cloaked scraps of hog slaughter,

phlegmy green pulp roped through relishes, jellies. Waiting until deep
winter, when our coal pile would dwindle and Jiggs in the funny papers
would seem suddenly not so funny. When the beauty of infinite variety
would shine through the brine and new virtues would reveal themselves.

— Nancy Kober

Bad Rap

Yo tomatoes
you're red and green
and you taste disgusting
with whipped cream!

— Jane Heblich, age 10

Yo tomatoes
when you get smashed
you really look
like you are trash!

— Isabel Heblich, age 7

Boing! Tomato

Once there were two tomatoes—a father and a son tomato. One day they were taking a walk. The little one was far behind. The father said, "Hurry up, or I will get very mad!" And the little tomato did not hurry up. So the father tomato went back to his son and banged him on the head and said, "Catch up!"

— *Heather Mariam Gross, age 7*

A Tomato I Once Knew

It talked then it stopped then
it rolled then it choked, it
took a deep breath, then
it walked the rope, after that
it croaked. A tomato I once
knew.

    It sang then it
swam, then it laughed,
then it laughed again, then
it left me. A tomato
I once knew.

— *Emily Morris, age 10*

A Tomato from Grandma

You let go my hand to bend over
and pick it up. You brushed off
the dust from the garden soil
where it had fallen from the vine.

It seemed small in your palm, Grandma.
You placed it big,
red and warm, in my cupped hands.

I held it to my nose,
breathing a sweet, dusty
summer smell.

Go on, you said,
Bite into it
like an apple.

I did
and you laughed knowing
I hadn't tasted it.

You wiped my chin, and the front of
my play dress with
a striped dishrag from your apron.

When you grow up, you said,
You'll use them in salads and sauce.
Or maybe,
you'll eat them
like apples
to ward off ills.

— *Blaise Zerega*

Identity Crisis

As I hang here and wonder,
As to which category I'm under,

Whether I'm vegetable or fruit,
I'd like to know the truth,

Fruits are sweet or nicely sour,
Vegetables are green and extremely dour,

Cherries, berries, apples and pears,
Of these genetics I'd like to share,

A brother to the brussel sprout!
The thought of it makes me wanna pout,

But a peach for a cousin,
Now that would keep me a-buzzin',

Hey?! I am in sauces, salads and juices,
I AM TOMATO! Over a hundred uses.

— *Al Stephens*

Family Garden

Winter's over and the snow is gone
Go get your gloves and pull your work boots on
Gotta clear that patch, pick them stones
The sun's gonna warm those lazy bones
Winter's over and the work's just startin'
Time to get busy in the family garden

Turnin' the soil and plantin' the seeds
Layin' down mulch and pullin' up weeds
Hoein' the corn, row on row
Then you stand right back and watch it grow
Even little children gotta do their part in
Helping things growin' in the family garden

Cool spring rain and summer breeze
Yellow squash and black eyed peas
Japanese kale and pinto beans
Italian tomatoes and turnip greens
Now we're picking and canning and working hard in
Trying to beat the winter to the family garden

Beans in the jar, potatoes in the bin
Squash in the cellar, winter in the wind
Even the youngest gardeners know
You're gonna reap just what you sow
Now all say grace and I beg your pardon
Won't you pass them pickles from the family garden
Berries on the bush and apples in the tree
Growing in the garden like you and me.

— John McCutcheon & Si Kahn
(Appalsongs & Joe Hill Music, ASCAP)

Caution: Pommes D'Amour

Though in the midst
Of our own garden
And glowed upon
By what we ourselves have grown
And fruitily glowing back,
We let doubt sprout with the love-apples:
To mate? O! O!

— *Stephen Margulies*

Getting from the Garden

Tom ate an o.
He loved all letters
But especially the o
Because he felt
Its fresh light fullness,
Its lively acidity,
Its bursting reality of redness,
Its alert love-juice.
O! It had skin popping-tight
As a balloon.
Just looking at it made him
Open his mouth
With knowing innocence.
(Adam and Eve's first true sound
Was, "O!"
Their mouth taught vowels
Through the taste of shape.)
Tom had pinched the o
Out from under a small sleepy
Dragon's green young wing
Which had got stuck in ground
Specially and richly caressed.
The dewy dragon did give up
Its tiny engine from Eden,
Its lovely acid living gem,
Its popping small heart.

— *Stephen Margulies*

Juice Atonement

Tomatoes are such harlots—
flaunting their bulging roundness
in skin-tight bright colors,
   O Lordy! have mercy
on those fuzzy spearmint vines.

Don't lay the seedful sinners
on your burgers and sandwiches
for they must repent—
baptize them on the stove eye.

Picante sauce is purified
marinara sauce sanctified,
and what could be more holy
than a good minestrone?
Where would we all be
without that altruistic ketchup?

The fruit group and the vegetables
may shun the ambiguous tomatoes,
but don't write them off as   bad.
They have a lot to offer   if
you bludgeon the bulbous orbs.

                    — *Tonya Miller*

Lust

Tendrils being as mere flourishes,
Climb and wind their way upward, always upward.
Imagine sprawling vines, leaves deep green
But with a dense down furring the surface,
Their undersides delicate and veined as wrists.
Summer's rush breeds rank weeds, extravagant colors
Of pomegranate, tomato, orchid; a heightened pulse
No one could resist.

Left behind in spring, virginal hearts
Still tick with seed's stored-up abandon.
What's vulnerable here remains unexposed,
And bare expanse of soil is testament
To ebbing possibility that dormant shoot
Will stir, forcing its way into new air.

— Kristen Staby Rembold

Ode To Tomatoes

The street
filled with tomatoes,
midday,
summer,
light is
halved
like
a
tomato,
its juice
runs
through the streets.
In December,
unabated,
the tomato
invades
the kitchen,
it enters at lunchtime,
takes
its ease
on countertops,
among glasses,
butter dishes,
blue saltcellars.
It sheds
its own light,
benign majesty.

Unfortunately, we must
murder it:
the knife
sinks
into living flesh,
red
viscera,
a cool
sun,
profound,
inexhaustible,
populates the salads
of Chile,
happily, it is wed
to the clear onion,
and to celebrate the union
we
pour
oil,
essential
child of the olive,
onto its halved hemispheres,
pepper
adds
its fragrance,
salt, its magnetism;
it is the wedding
of the day,
parsley
hoists
its flag,

potatoes
bubble vigorously,
the aroma
of the roast
knocks
at the door,
it's time!
come on!
and, on
the table, at the midpont
of summer,
the tomato,
star of earth
recurrent
and fertile
star,
displays
its convolutions,
its canals,
its remarkable amplitude
and abundance,
no pit,
no husk,
no leaves or thorns,
the tomato offers
its gift
of fiery color
and cool completeness.

— *Pablo Neruda*

The Okra Flower

I stood in a ripe twilight
meaning to think about the mountains,
their brilliant hem all around,

but thought instead, I do not want
to die here, away from home, away
from her as she goes to the garden
to gather in this long light,
as she breathes the tang of tomatoes
and feels her forearms prickle
when she stretches to cut the okra.

I remembered how the okra flower
would be folded for the night,
how it held in itself the colors
of her face better than any picture,
the moon of her skin,
the rich purse of her lips.

I thought, when I get home again
I will stand at the kitchen window
and watch her stitch up the beans,
and a kind of healing will begin,
until the days ripen like a row
of vegetables on the bright sill,

until I can walk in the garden
on an early September afternoon
and look deep into the okra flower
without smell, without a freckle,
and not think of her, not her.

— *Michael McFee*

Adolescence—III

With Dad gone, Mom and I worked
The dusky rows of tomatoes.
As they glowed orange in sunlight
And rotted in shadow, I too
Grew orange and softer, swelling out
Starched cotton slips.

The texture of twilight made me think of
Lengths of Dotted Swiss. In my room
I wrapped scarred knees in dresses
That once went to big-band dances;
I baptized my earlobes with rosewater.
Along the window-sill, the lipstick stubs
Glittered in their steel shells.

Looking out at the rows of clay
And chicken manure, I dreamed how it would happen:
He would meet me by the blue spruce,
A carnation over his heart, saying
"I have come for you, Madam;
I have loved you in my dreams."
At his touch, the scabs would fall away.
Over his shoulder, I see my father coming toward us:
He carried his tears in a bowl,
And blood hangs in the pine-soaked air.

– Rita Dove

Farm Poems

> for my Grandmother

IRRIGATION

Halfway down the back steps
with a pan of cobs
and oyster shucks,
I am startled
by a disturbance in the corn rows:
colossal silver birds
swing heavy wings up over the corn beds
and reel.

TOMATOES

We peeled them
right out of bushel baskets.
Squatting under sycamores
with knives and a basin,
Grandmother, you taught me
to tug the skin gently
from the warm meat,
to cup the heart
with its smooth, protruding
vessels, its seepage,
to my ear
to hear the suck and purl
suck and purl.

— Lisa Russ Spaar

All Tomatoes

Unlike potatoes they have
no need for eyes
growing as they do
in the eye of God

their only threat
the caterpillar
lumping its way
up the vine

or the bird
overhead
eyeing the first
rosy blush.

In their kingdom
after planting
come sun, water, growth
then the ritual

of teeth, tongue, taste—
the yearly spilling
of juices
their sacrificial gift.

— *Connie J. Green*

After Too Much Sanity

It must have been something about the hush
of the room—the screen and flashes of color
disturbing the darkness—that drew the owl
to the ledge outside my window.
If glass hadn't intervened,
I could have touched him, smoothed
the ruff of feathers around his enormous
profiled head. Then the face turned backward.
His eyes held mine. Like a lover who knows
everything you haven't told, the bird,
blinkless, made me uneasy, warm,
so that if someone had entered the room,
I'd have blushed, startled
from indiscretion. The way the tomatoes,
still green, yet plump with juices, seemed
last summer—not on their stems, where they should
have continued ripening, but on the ground
where the squirrels rolled them, knocking them
from the tops of six-foot high stalks—
too brightly intimate. Bloodless,
they lay in the open, the grass beneath them
secretly glad for the weight. I saw
the way they shaded into each other
so that, in certain light, one could hardly tell
globe from blade. I stood at the kitchen window
and watched as the last potential lusciousness
landed. Swore my disaffinity with innocence.
Flushed with skewed desire.

– Susan Ludvingson

Harvest

We were eating tomatoes in the wet heat
of August, passing them between us like apples,
juice dripping down our chins, staining our shirts.
"Come get your red hot tomatoes," I said,
licking my lips. You sighed, said, "I planted these
with my true love last spring. Watered and weeded,
but we barely made it to the spinach and chives.
She left behind budding broccoli, zucchini
the size of your finger." You waved a tomato
in the air, reached across the table for my hand.
"Planting a garden with your lover is the kiss of death,"
you said. I laughed, took the last bite. We had just met,
had planted nothing, so were safe for the season.

In fall you called to say you'd stripped
the vines—early frost coming—and there were
scores of tomatoes, tomatoes
to burn. Beyond you I heard pots bubbling,
reducing the fruit to sauce. You brought me some
green ones, said to wrap them in paper, store
in a warm dry place, soon enough they would ripen.
Then winter, and blank snow, and for weeks
not a word, nothing from you, but the dream
tank's full up, it's overflowing, as if
our daily silence fueled this voluptuous nightlife
of the soul: crumbling clay towers, sea-green oceans,
full moon crooning of cutworms and love. And
all along the tomatoes turning, disappearing one by one.

— *Sydney Blair*

Putting up Tomatoes

In the steamy kitchen, the women peel tomatoes,
murmur the news of the day: an accident,
an unpaid bill, a woman losing her mind.

She leans in the doorway
newly widowed, new teeth clicking
with each syllable she speaks
"Turn them beauties gently."

These are her husband's sisters,
five stout women
pink with the noonday sun and the steam,
come to fill the clean glass jars
with the round, sunred, apples of love.

On the porch she studies the garden
where he plowed
each year, his back bent and brown
as she held him
still sunwarm at night.

— *Amy Spence*

Okra

Petals, six, lapped
like the armadillo's
or the pine cone's scales,
their color primrose, yellow
as first sun on something
white, fence paling, church
portal. Its blunt pistil
black-tipped, thick; the pistil's
bed, base of the flower cup,
a blotch, insistent, black
and russet velvet. Calyx
crumpled like a silk dress
dropped.

The fruit comes downy,
pushing its little point up and out,
green as anything young,
fern, milkweed, asparagus,
and like them, best taken
early, its planes and ridges
dew-crisp. Have it raw,
or brought just to heat
if the taste is new, or,
if you know and love it,
bathed in tomato blood
and its own rheumy glue,
pod relaxed, pearly little seeds
on the loose.

— *Shirley Anders*

Poor Song

Can't abide; can't quite die—
Wish I could sigh.
Just a wifeless piddling beast
Looking out for milk or meat:
Forgot my nametag ages ago—
Once I computed
Or maybe disputed:
Now I'm, hell,
O, really, well,
Without a game.

I must get a beggar's cup,
You must get a beggar's cup:
In a white cafe
That the moon has scrubbed,
On catsup and water we'll sup, we'll sup;
On catsup and water we'll sup.

— *Stephen Margulies*

The Manikin Leaves

Through smashed plate glass, the sudden smell
   of diamonds and cashews, orange peels, tobacco-
     juice and piss, and a tortoise-shell cat,

off on night whims, rubs liquid-electric
   against her legs. The manikin's shimmied off
     the steel rod, sick of air conditioned

to 68 degrees, silent faces and silent street.
   A strange ticking begins in her body. Rain
     has glossed the lamppost, a furry dogwood

leaf, and her reflection looks up, cracked
   from the pavement. Behind her the boutique owner
     shouts *You have a few good years*

left to her heap of clothes in the window.
   She went on some time before the Tin Man,
     dancing from behind a lamppost, romanced her

with his tin arms, pulsing with tiny lights.
   He wakes her in his only room at midnight
     with stew bubbling in a can. But she must find

her own opinion—his tin face roaring
   up the fireplace of her thighs. She walks
     the streets pushing her face into strawberries,

plunking lettuce on her breast and sprawls
    on a heap of papery-white onions; they roll
        in concentric circles, like a pile of pebbles

hit with a fresh pond. The vendor is indulgent;
    she is so new. A sassy jay
        is in the vacant lot. The ruddy odor

of tomatoes from a sidewalk cafe
    where a pigeon coos under a table.
        She flings it a loaf of French bread

as she writes her shopkeeper a memo: *good bye,*
    and a safe retirement. She could only
        be glad for so much unpredictable air.

                    — *Jane Caris*

For My Daughter

I stand between rows for hours, fasten loose ends
to stakes with old pieces of cloth,
try to remember what I must tell you: Never swim alone,
don't burn in the sun, beware of strangers on the beach,
don't leave me . . .
After they come for you I return to the garden,
avoiding the house, drained of your presence as I am.

Watching the pool bottom slope out of sight,
I swim faster. Yesterday you called, told me sharks
migrated past during the night. Newspapers
were full of it, offered good advice:
Don't swim with a horse or dog, never molest a shark.
We laughed and you were gone.
Today I am swimming alone, the only predator
my shadow bumping against the concrete walls of the pool.

After you hang up, surf still pounds in my ear,
you return to the beach, poised
between the woman you will become, the child you were.
I sleepwalk into waves, wake between borders
where tomatoes soften in the sun, my fingers closed
around one. Words run in my head like newsprint under water.

Today I will not think of you. Heat rises in the kitchen
where I drop tomatoes into white enamel pots
hissing on the stove. They surface with skin split,
peeling from the red core, bleed into clear plastic;
on freezer shelves ice crystals form,
pink as froth on the lips of the drowned.

— *Judy Longley*

To Keith
(1954 - 1972)

In my dream you're waiting on the street,
idling outside the Lariat Cafe while I
count my tips and put the ketchups away—
you're leaning out of that
jacked-up pink chevy we'd bolt town in
to tune in the rock from Oklahoma City
and skim along moon-dim back roads,
going nowhere. You raise your arm to me,
the one with the hand cut off, though that's not
the way you did it. You pulled the chevy
into the YOU-WASH-IT after school,
rolled both hinged doors down to the concrete
and slid back in behind the wheel,
radio up, motor roaring.
But in the dream you're still waiting for me,
your old girlfriend, the one
you gave your diaries to that day
—and I didn't ask why, though we weren't
going together anymore; I didn't ask,
didn't even look before
I stuffed them in my locker at lunch;
I liked being the keeper of feelings,
being needed but not touched.

— *Debra Nystrom*

Cancion de Segadores

       In the thick adobe
harvest room
back of the old family house
he watches
his sister gather
into bushel baskets
caps and whiskey bottles.
From his bed he hears
her children playing outside,
probably priming the old pump
beside the cottonwood
later they will climb.
Sometimes
    they seem far away.
      Now
on the front porch
they suck honey from honeysuckle
into their voices.
      Once
he and their mother
    skinny and brown
    small in that sun
stood on tiptoe
to look over the adobe garden wall
at their grandfather's desert
blooming.

    She places the moistened cloth
on his forehead, he sees
her look and look away.
    The children
play in the main house, talking on the
dead black phone.

He was nine when he forgot
to close the corn box. Next day
his father found the swollen
        horse.

        And
her *hijita,* too little
to have known the garden and horses,
finds the piano.
                Notes rise and rise
and fall again.

She named her Felix for him
who would peek
from behind her skirt
at the Tío with his name

        In the field
where the children's voices
rise through afternoon dust,
        Felix walked
with his Tío between
        rows of corn.
        In green dawns
Tío would pick an ear,
pierce the kernels, let Felix
suck in the milky sweetness.

        He learned to sweep
his fingers through tomato plants
to hold fragrant onto summer,
and to pick squash blossoms
before the bees become entrapped
          para que las frie tu Mamá Margarita.

        At the acequia
they would stop,
open the water gates.
        He would watch the
furrows fill with brown foam
and breathe
        the morning Río Grande.

        The children's voices rise
a tisket a tasket
a green and yellow basket

        and fade
as he remembers skipping with his sister
beside Mamá Margarita on the Sunday
way to San Felipe
inventing verses
          Yerba buena para la fiebre
          Yerba buena para la fiebre
          más alta la fiebre
          más fuerte la té

        Beside his bed
she bends,
offering a bowl of steaming tea
Yerba buena para la fiebre

He closes his eyes.

In the thick-walled adobe silence
of his father's harvest room
he sees
red chile ristras
hanging over
        garlic, onions, squash,
        baskets of corn, pumpkins
        potatoes in bushels,
        tomatoes in jars and
        San Felipe parish
        apples, peaches, pears,
        young wines in battered barrels
        and sacks and sacks of
            mountain pinto beans.

                      — *Antonia Quintana Pigno*

Ephemeral Landscapes

Los Angeles twists out of the desert
& the stars fall, dominoes in the lithic dark.

Right now my sister wades out of the sea, the salt
like chainmail on her legs. Something we

talked about when we were driving here: I do not
remember my sister as a child. She has no memory of me.

We exist inside each other's minds and early lives
the way the orobouros consumes its own tail—

the silk of the scales, the heat & bitter edges.
The soul, I think, wanders around inside the body

like a ravening animal. It snuffles among velvet vines
of wild tomatoes, blinks dumbly at the sun.

Where, then, is memory? Where am I, where is she?
Here, now, I am lucid & recalcitrant. I have no questions

yet my sister urges me to faith, to that ephemeral
landscape of desire, of perfect & unwitting servitude.

Your life can change in an instant, she says.
She wants me to believe the infusion of grace,

that the Holy Spirit can reside in my body
like a traveler at a motor lodge. But my body

is crowded with history and when I think of Jesus
I think of Mary, only Mary. Her body was also inhabited,

conquered like a vast territory & named acquiescence
by diligent cartographers. —Quietude, the scent of roses:

Close by shore the fathom trawlers tie up for the night.

— E. A. Gehman

A Letter to Michael McFee: Italian Tomatoes

My gift came at an odd time, not wrong,
just odd, plants ready to be dropped
into the soil of your garden
that will not bear till you are gone
north. A good thing Belinda will be here
to gather the small fruits, their cool bodies
luscious Renaissance nudes, long-spined,
hips smooth as porcelain. She will save some
for you to see. The seeds
came from Venice. The idea teases us.

On the way home from our afternoon
together, I thought of you and,
sentimentally, how things come to fruition
at their right time, early or late; the odd
time, the odd place: Ithaca, or where
dead writers—Jarrell, Lettie Rogers—
walked the bricks with little
animosities and big ideas. Most of us
need to move on, sometime or other,
if only later to be able
to say the obvious, "Here.
I came back. It's all right.
This is all right. This is pretty nice."

These tomatoes will make good sauce.
You won't mistake them, use them for salads,
sandwiches; for those you'll grow
other kinds, Early Girl, Better Boy. Don't
look for anything remarkable, these aren't
rare, you could match the seeds at Burpee,
easy. But because they come from me,
from my friend a mother of six who teaches
English, before her from Pound's city, to you
they'll be special. That's what pedigree
is, what legacy is, I suppose:
the history of the seed.

— Shirley Anders

The Nature of Yearning
for Lindsey and Bess

I
This Northern August swells with warmth
the garden would burst and a trout waits
beneath the moving river surface
he holds steady until the brown caddis
fly floats above him he plunges upward
breaks the silent water then slaps down fat
as deer that graze the flat meadows while slowly
as in a dream of shadows a black bear circles
beneath trees ten thousand shades of green.

II
All changed now quickened the morning
air in September lifts the spirit high
as one perfect trumpet note still
this clarity this concerto suggests
the coming death of tomato vines also
cucumber broccoli corn beans peas yellow
squash cauliflower all vegetables dead
or dying we wait the swelling of pumpkins
the blood flame turning of delicate leaves.

III
High down out of Canada the geese were flying
all day my wife said that far-off honking
sound makes you feel lonely the trees were pure
fire for two weeks but now the leaves have fallen
all purple and brown the woods resound with axes
while men cut logs the children home from school
go out for kindling the leaves crackle the blood
of the animals flows richer and a white-tail doe
sniffs the air at dusk her smoky fawn now half-grown.

IV
Chop the caught turkey's neck catch the buck
deer in gunsights fire shots deep into his heart
sling up his carcass to a thick tree cut open
his belly and handle the bloody heat and stink
of his guts shoot doves partridge quail
pheasant and grouse shoot rabbits shoot quick
squirrels and walk the stubbled fields with meat
on your back for soon the snow comes and with it
the silence at night when the wind wants man flesh.

V
White December the elegance of pine trees
in snow with voices rising in praise of Christ
the soft child of winter all Bach Fasch
and Handel cannot hold Jesus' swelling song
but now the trout takes no food the bass
sinks into the darkest pools of the river
the bear's blood slows while goose and duck
have long flown south and beside the house
snow deepens over logs stacked for the fire.

VI
Ice ice the death of trees the wind strips them bare
it whips them into savage rooted dances branches
crack limbs are yanked off they fall and smatter
on the frozen ground fearing wind I tell my wife
don't stand by that window a pane might burst
this morning she found stiff on the crusty ice
a redpoll dead and light as dust in her hand she said
the sun has forgotten us the nights go on and on
the clouds flee and the wind howls all day long.

VII
No meat in the house we cut holes in the ice
this February we fished for smelt and perch the ice
on the lake was two feet thick my wife thinks
the birds have left us forever only the rats thrive
they steal our corn and leave us just cobs and husks
rabbits are hard to track now but one the other day
sat in the field he was so cold I walked up and kicked
him before I shot the ice builds its kingdom
and holds against what fire we have left love.

VIII
We long for warmth these days there is little sun
still no birds have flown north over our house
and I think this March no month for birth only
the wind has life no green anywhere the trees
are just bones they shiver and bend they want
loose from this earth yesterday we saw the grass
it was brown and dead as an old hide in daylight
the snow melts some but it freezes again at night
the ground is covered with brittle crusts of thin ice.

IX
Oh the waters burst there are the timid green buds
delicate grass crocus and daffodil the waters
gather they flow out of the mountain the streams
wash off dead limbs and leaves the gentle rains
bring birth this air of April wakes even the animals
the spring birds have come back the trout leaps again
now the wind is a child the earth is sunlight a woman
walks outside this morning she is beautiful as the clear
sweet sound a man makes with his horn at his lips.

— David Huddle

Cash Crop

Money skimpy as the skirt of your one good suit,
you weigh seed in each palm until tomatoes sprout
green, rank along the suburban street. At bird dawn

in sensible shoes, you labor between rows
until the sun runs down its ration of gold, nip
new growth where foliage threatens fruit,

succor with hose, city water
afternoons cloud sweep past, their hoard of rain
unrelinquished. Once bulbs shine

among dark leaves, you count and spray, cutworms,
slugs, a wealth of pale corpses behind you.
I think you a child, your tired face a gleam

between sunset and bed, naming things,
things you'll buy if the market holds. I want to say,
Mother, remember the chickens. A thousand

crowded into the shed, light bulbs strung
so constellations circled each miniature sun,
sprouted feathers, grew to frying size,

sold for less than feed and energy. Still you dream
their pale eyes, heads bobbing from crates,
a white plume settling on the curb as their truck

rounds our corner. When you sleep shoes stand guard
from the back porch, down at heel,
caked mud drying. If they could sing

it would not be the whippoorwill's idle call
but a song to ease your laborer's stoop,
that hope like labor is its own reward

as the scent of green mingles with damp earth,
swells from garden to house, sets dreams ticking
toward the future, its rich, ripe heart.

— Judy Longley

Contributors' Notes

Debra Allbery's book of poetry *Walking Distance* was the 1990 winner of the Agness Lynch Starrett Poetry Prize. She is the recipient of fellowships from the National Endowment for the Arts and the New Hampshire State Council on the Arts, and also won *The Nation's* 1989 Discovery Prize.

Shirley Anders is a displaced North Carolinian who teaches English in Wisconsin. She is the author of *The Bus Home,* winner of the Devins Award for 1986.

Sydney Blair administers the creative writing program at the University of Virginia. Her first novel, *Buffalo,* won the 1991 Virginia Prize for Fiction. She is currently finishing her second novel, *Jan Gets a Dog.*

Jane Caris directs the counseling center at Guilford College in Greensboro, N.C., where she lives in a farmhouse and her hand-built mountain cabin. Her poetry has appeared in *Rhino, Kentucky Review* and *Poetry Miscellany.*

Rita Dove is the poet laureate of the United States. A professor at the University of Virginia, her books include a novel, *Through the Ivory Gate,* a book of stories, *Fifth Sunday,* and the poetry collections: *The Yellow House on the Corner, Grace Notes, Museum,* and *Thomas and Beulah,* which won the Pulitzer Prize in 1987.

Helen Fahrbach's two books of poetry are *No One Rides The Carousel* and *A Thousand Journeys.* She is a retired librarian and an avid gardener who lives in Neenah, Wisconsin.

E. A. Gehman is a video artist and poet who lives in Lynchburg, Virginia, where she is an assistant professor of English at Randolph-Macon Woman's College. She was a Hoyns Fellow at the University of Virginia and a Xerox Fellow at Hollins College.

Connie Green, a weekly newspaper columnist from Lenoir City, Tennessee, is the author of two novels for young people, *Emmy* and *The War at Home.* Her poetry has appeared in *Confrontation* and *Cumberland Poetry Review.*

Heather Miriam Gross recently graduated from the second grade. She loves cats and math and dancing.

Jane and Isabel Heblich don't like tomatoes but do like poetry. Their mother is the editor of this book.

David Huddle's books of poetry include *Paper Boy* and *Stopping by Home* and his most recent short story collection is *Intimates*. A native of Ivanhoe, Virginia, he now lives in Vermont and teaches at the University of Vermont and the Bread Loaf School of English.

Si Kahn is a songwriter and community organizer from Charlotte, North Carolina. He is the executive director of Grassroots Leadership and records for Rounder Records.

Marilyn Kallet is Director of Creative Writing at the University of Tennessee, Knoxville. Her most recent book is *A House of Gathering: Poets on May Sarton's Poetry* from The University of Tennessee Press.

Nancy Kober received her M.F.A. in fiction from the University of Virginia and lives in Charlottesville. Her fiction and poetry have appeared in *Sundog, Timbuktu, The Kennesaw Review,* and *The Writer's Eye*.

Ellen Kort of Appleton, Wisconsin, is the author of seven books and a play and the winner of *Nimrod* magazine's Pablo Neruda Prize for Poetry. Her poetry is included in two recent anthologies, *Inheriting the Land, Contemporary Voices from the Midwest* and *If I had My Life to Live Over I'd Pick More Daisies*.

Maxine Kumin was awarded the Pulitzer Prize in 1973 for *Up Country: Poems of New England*. Author of a dozen books including *Halfway, The Retrieval System,* and *The Abduction,* she lives on a farm in New Hampshire with her husband. She's also a winner of *Poetry* magazine's Eunice Tietjens Memorial Prize.

Judy Longley's book *My Journey Toward You* is the recipient of the Marianne Moore Prize for Poetry and will be published by Helicon Nine Press. Her two chapbooks are *Parallel Lives* and *Rowing Past Eden,* and she is the poetry editor of *Iris: A Journal about Women* in Charlottesville, where she works as a substance abuse counselor and therapist.

Stephen Marguiles is curator at the Bailey Art Museum at the University of Virginia, an art reviewer and a literary reviewer and did not realize how much tomatoes meant to him until the editor's suggestion aroused bittersweet memories.

John McCutcheon is an international folk musician whose latest album is "Family Garden," his third collection for children and families, following his award-winning albums "Mail Myself to You" and "Howjadoo." He lives in Charlottesville with his wife and two sons.

Michael McFee was born and raised in Asheville, N.C., but has been growing tomatoes in Durham, N.C., his wife's hometown, for 14 years. He has published four books of poems and teaches poetry and writing at the University of North Carolina, Chapel Hill.

Rusty McKenzie is an art therapist and poet who lives and writes on the north shore of Lake Winnebago and sometimes on a tiny Bahamian out-island. Her poems have been published in *Wisconsin Review, Rag Mag, Fox Cry,* and *Columbia Review.*

Tonya Miller graduated from the writing program at the University of Tennessee with honors. She has completed a book of poems called *The Soul's Talk.*

Laurel Mills teaches creative writing workshops for the University of Wisconsin. She is the author of three collections of poetry and her poems have appeared in *Calyx, Kaliope and Yankee.*

Emily Morris is a fourth grader in Knoxville, Tennessee. She is a dancer who performed the Nutcracker with City Ballet of Knoxville, a painter and a writer of short stories.

Debra Nystrom was awarded the Virginia Commission for the Arts Prize for Poetry in 1987 and the *Virginia Quarterly Review* Balch Prize in 1991. Author of a book of poetry called *A Quarter Turn,* she teaches English at the University of Virginia.

Antonia Quintana Pigno, a native of New Mexico, teaches Spanish in Kansas State University's Modern Languages Department. Her two books of poetry are *Old Town Bridge* and *La Jornada.* She reports that tomatoes from the desert are always remarkably sweet and wonderful.

Kristen Staby Rembold of Charlottesville is a poet and fiction writer whose work is forthcoming or has appeared in *Nimrod, Phoebe, Iowa Woman, Appalachia* and other magazines. Her chapbook, *Coming into This World,* was published by Hot Pepper Press in 1992. She serves as fiction editor of *IRIS: A Journal About Women.*

Nila Jean Robinson is a criminal defense attorney in Appleton, Wisconsin. Her poems have appeared in numerous Midwestern publications.

Bobby Caudle Rogers was a Hoyns Fellow at the University of Virginia, where he received an M.F.A. degree. He's an assistant professor of English at Union University in Memphis and his work has appeared in *Southern Review, Georgia Review,* and *Shenandoah.*

Joan Z. Rough is a photographer whose work has been on exhibit at the McGuffey Art Center in Charlottesville, as well as other galleries. She's also the author of a book on the lost art of Australian locker hooking, and her poetry has appeared in various journals including *Life Scribes.*

Enid Shomer's poems and stories have appeared in the *New Yorker, Poetry,* and *Paris Review.* Her newest books are *This Close to the Earth* (poems) and *Imaginary Men,* winner of the 1993 Iowa Short Fiction Award.

Lisa Russ Spaar's family has grown tomatoes for three generations. Her grandfather, his father and his father's father sold tomatoes to Campbell Soup in southern New Jersey.

Amy Spence, a candidate for an M.F.A. degree at the University of Virginia, works in both poetry and fiction. She lives with her husband and two children in Charlottesville, Virginia.

William Stafford's latest book of poetry, *My Name is William Tell,* won the Western State Lifetime Achievement Award in Poetry. He's won a National Book Award, the Award in Literature of the American Academy of Arts and Letters and the Shelley Memorial Award. He has served as a consultant in poetry for the Library of Congress.

Mariflo Stephens writes fiction, essays, and poetry from her Charlottesville home. Her work has appeared in various publications including the *Virginia Quarterly Review, Iowa Woman, Catalyst* and *The Washington Post.*

Al Stephens, a dairy farmer in Wytheville, Virginia, is a 1990 graduate of Virginia Polytechnic University with a B.S. in Dairy Science. He was a winner in the short story division of a Chautauqua Creative Writing Contest.

Blaise Zerega lives in San Francisco where he writes poetry and fiction. An avid fly fisherman, he studied writing in Charlottesville, Virginia.